Arthur Smith

The Past, Present, and Future of Technical Education

In Connection with Agriculture

Arthur Smith

The Past, Present, and Future of Technical Education
In Connection with Agriculture

ISBN/EAN: 9783744669818

Printed in Europe, USA, Canada, Australia, Japan

Cover: Foto ©berggeist007 / pixelio.de

More available books at **www.hansebooks.com**

THE

Past, Present, and Future,

OF

TECHNICAL EDUCATION

IN CONNECTION WITH

AGRICULTURE.

BY

ARTHUR SMITH.

ONE SHILLING.

St. Albans:

Printed and Published by Gibbs & Bamforth, Market Place.

1891.

PREFACE.

IN preparing this little work, the writer has endeavoured to avoid mistakes either in statements of fact or statistics, but at the same time errors may possibly have crept in, and in the latter case he would esteem it a favour to have them pointed out. The information in Appendix A. has for the most part been kindly supplied by the clerks and organising secretaries to the various County Councils, to whom, for their courtesy in supplying, the author takes this opportunity of tendering his best thanks. For the matter in the Appendix containing the list of places where lectures in agriculture were given last year in connection with the Science and Art Department, the writer is under obligation to Professor Buckmaster. The statistics, etc., relating to State aid to agricultural education abroad, have been taken from the Blue Book, containing the report of the Royal Commission on Technical Instruction.

ARTHUR SMITH.

SMALLFORD, ST. ALBANS,
December, 1891.

CONTENTS.

The Past, Present, and Future of Technical Education in connection with Agriculture in the United Kingdom.

INTRODUCTION.

Among the multifarious methods adopted by man as a means of livelihood, none can be said to at all approach agriculture in importance.

(margin: Agriculture the most important industry.)

All trade and commerce rest on or commence by the primitive activity of the farmer. It is, as Emerson has well put it, " his part to create wealth." But there has been a great reaction, even for this period of rapid change, since the days when this essayist wrote concerning husbandry, that it was not only a physically exhilarating pursuit, but an unfailingly profitable one also. Emerson advised those who had wrecked their capital in town occupations to go back to the land and recruit. However excellent this counsel may have been when given, there are few, if any, who would act upon it to-day. Farming is now chiefly remarkable as being an insatiable drain upon one's exchequer. To enter into all the real or fancied causes which have given rise to, what has unfortunately become an everyday expression, "the agricultural depression," would not be germane to the present subject ; still less would it be pertinent to discuss the immense variety of so-called remedies which have from time to time been put forward as being calculated to alleviate it. There is, however, among the latter, one, viz., protection, about which a word *en passant* may not be out of place. Amid the many quack specifics none have had a more baneful effect than this, because the idea, which too many of the agricultural interest appear still to entertain, that a return to protection is possible, has rendered them blind to helps which are within their reach ; thus losing the substance while running after a shadow, and in this way retarding more than hastening the realisation of the end that they have in view.

Probably one of the most potent causes of failure in connection with agriculture is, that the rank and file of this interest take up their profession with scarcely any previous preparation.

(margin: Want of professional training, one of the causes of failure to make farming pay.)

In some cases the sons of farmers, after leaving, at the age of 15 or 16, an ordinary middle-class school where " home comforts" are the most prominent feature in the prospectus, are put to farm labour. If they are kept steadily at it and are made proficient in every kind of work performed on a farm, it is a good professional training as far as it goes. The more common way, however, in which these youths spend this period of life, consists in loitering about without any stated occupation, attending fairs and markets, the race course and the hunting field, which is about the most absurd and pernicious that can be imagined. Such young men are truly to be pitied, for they are neither inured to physical labour, nor afforded the inestimable benefit of being practically, still less scientifically, educated for their calling in life. It need not surprise any one that such hapless lads should be incompetent for making farming pay. The idea of special education being of any practical value to the farmer has been treated by many as a palpable absurdity ; yet of all the professions none can gain more benefit from it than agriculture. It is certain that in the future the farmer who has been technically educated will take the lead. The merely practical man whose mind can only hold a few ideas, will give place to one who, while thoroughly well versed in every practical detail of work and management, is at the same time a man of education and scientific skill. All farming operations are capable of being explained on scientific principles ; and no one can be said to understand farm-work who is unacquainted with the teachings of modern scientific research. What is really required is a knowledge of both the practice and theory of the art of agriculture, to enable a man to cope with the many difficult, but not insurmountable, questions, and circumstances which are incidental to this industry at the present time. To impart this, in a thoroughly efficient manner, is the object of technical education. By itself practical knowledge is largely of an empirical character.

Experience may have shewn the farmers of a particular locality in a course of years what style of work will enable them to " get along somehow " without knowing the why and the wherefore of a single operation. But immediately they move to a new locality, or state of markets and trade arise which necessitates a change in practice, they are quite at a loss how to proceed and often commit ruinous blunders. Experience has taught them what to do under a given set of circumstances, directly those circumstances are altered they are thrown " out of harmony with their environment," and it is a chance whether they fail or succeed. Even those who are comparatively successful, practical men, would be still more successful, if on to their practice there were grafted a knowledge of science. The late Mr. J. C. Morton, in a lecture delivered before the Royal Agricultural Society of England in 1865, said : " Agriculture, though not as a whole a science, has at length become a museum, as it were, of facts and specimens, in the classification of which students of all the sciences have been successfully at work, so that every part has now the light upon it of well-defined relationship with scientific truth."

During many centuries it has been imagined that agriculture, existing everywhere, might be exercised without any special instruction ; everyone, even the most dull in capacity, has been considered suitable for agricultural pursuits. Still, at the same time, there have not been wanting men with minds of advanced thought, who caught glimpses of the progress to be realised, although their field of action was limited, to point out the fact that the farmer requires something more than mere rule of thumb knowledge.

The art of agriculture has been studied by the greatest men among the ancients, and had been treated by the most learned and celebrated of their authors. Mago, a famous Carthaginian general, wrote twenty-eight books upon the subject; then we have Virgil's Georgics, followed by the writings of Xenophon, Aristotle, Theophrastus, and others. The first Latin writer on agriculture was Cato who was succeeded in this branch of literature by Varro, Columella, &c. Then we come to the *Mussi dominici* of Charlemagne. Later on Bernard de Palissy wrote: "There is no art in the world which requires greater philosophy than agriculture." In our own country we had, amongst others, Fitzherbert and Jethro Tull. All these were, however, only isolated efforts, the benefits of which could not spread far. At the same time they are worthy of note, as shewing, in the most eminent degree, that the greatest thinkers of the different periods, men of the strongest calibre of mind, were fully persuaded that the best educated man would make the most successful husbandman. Of course none of these writers had much comprehension of agricultural science as we understand it to-day; this was reserved for more modern times.

The first step towards placing agriculture on a more satis-factory footing was made by Sir H. Davy when in 1813 he published his "*Elements of Agricultural Chemistry.*"

This good work was advanced by the researches of Johnston, Boussingault, and Leibig. The latter, in a book published in 1859, says :—

"Agriculture, of all industrial pursuits, is the richest in facts, and the poorest in their comprehension. Perfect agriculture is the true foundation of all trade and industry. But a rational system of agriculture cannot be formed without the application of scientific principles." These pioneers, who laid the founda-tion of agricultural science, accomplished that which will benefit humanity to an extent which even to-day can hardly be realised. Unfortunately, however, agriculture has not, up to the present, made all the use she might of the labours of these and other scientific men. In this, farmers should perhaps be pitied rather than blamed; they have had in the past few opportunities of obtaining (except at a great cost) technical education. Even had the chance been given them, it is doubtful whether much good would have been accomplished, owing to the prejudice existing against science in the minds of the majority of agriculturists. But there are signs that this feeling is gradually wearing away; and in the future this period of depression will be pointed to as being not an unmixed evil, as it had the effect of causing the light of science to be allowed to shed its rays to the fullest extent on agricultural practice; thereby placing this profession on a firmer and surer basis than it ever before occupied, and elevating it, in the truest sense of the term, to the position of an art.

Probably at no period were the majority of farmers more ready to give attention to anything which promises to help them farm with better results, than they are to-day. It is, therefore, a very opportune moment at which to bring this question of a agricultural technical education before the country.

In discussing this subject, it will be convenient to first point out what have been the facilities for acquiring an education, in the highest sense of the term, to thoroughly fit the embryo farmer for successfully carrying on his profession, together with a brief description of the existing institutions, etc., which have been founded for the avowed purpose of providing this kind of instruction ; and then to proceed to bring forward suggestions for extending and improving the present system so that it may become available to the whole body of those who depend upon the cultivation of the soil for a livelihood, including the agricultural labourer.

The Past and Present. In giving a short statement regarding the principal agricultural colleges, schools, lectureships, etc., which are either in existence or which have become defunct, the dates of their foundation will be stated ; so that the opportunities, in the past as well as at the present time, which have been within the reach of those requiring the most advanced agricultural knowledge, will be made apparent. The description of these institutions is an easy task ; but an estimate of their value is more difficult. Opinions differ very greatly on nearly all educational questions, but when technical education is involved, and when its consideration is mixed up with general instruction, the diversities of ideas met with appear almost hopelessly irreconcilable. The difficulties in the case of an ordinary industry are increased enormously in the case of agriculture, for a reason which is peculiar to the cultivation of the soil. A bootmaker, weaver, ironworker, and, in fact, any other artisan can pursue every department of his calling at any time, and can teach an apprentice how to perform all the necessary operations every day in every week of the year. The teaching of agriculture means instruction how to perform a long series of operations, each of which can only be accomplished during a limited period once or twice in the course of the year. If, therefore, an agricultural student fails to grasp how, for instance, to prepare a stubble for the succeeding crop, he and his teacher must be content to wait nearly twelve months before the lesson can be resumed. This peculiar difficulty pertaining to the more practical part of agricultural technical instruction, is too often lost sight of by those who argue solely from first principles ; as well as by those who do not distinguish sufficiently between the kind and amount of technical education necessary for the landowner, the tenant-farmer or occupying owner, the farm-bailiff, and labourer respectively.

In a review of what has been attempted and acccomplished to promote agricultural technical education in the United

Kingdom, it will be better perhaps to take England and Scotland first, treating Ireland separately, because the circumstances connected with this subject in the latter differ essentially from those relating to the other parts of the British Isles.

In England no institution of this kind existed until 1845, in which year the Royal Agricultural College at Cirencester was established. This college owes its origin to an address given by the late Mr. R. J. Brown to the Cirencester and Fairford Farmers' Club at a meeting held in November, 1842, on the "Advantages of a Specific Education for Agricultural Pursuits." The discussion resulted in a resolution in favour of founding an institution where this kind of education could be obtained, being carried. A public meeting followed, resulting in the formation of a committee. By the persevering efforts of this committee and Mr. Brown a subscription of £12,000 was raised, principally among the landowners in various parts of the kingdom. For the first twenty years the college was not a financial success, continued calls having to be made for subscriptions and donations. This was, no doubt, owing to the low fees charged, in-students only paying £30 per annum. Since 1865 alterations in the mode of management have been made, which, together with an increase in the fees, has made the college financially and, for those who can afford the fees, educationally a success.

The college buildings include a museum, lecture theatre, class rooms, laboratories, dormitories, and apartments for the resident professors. As part of the establishment, mention should be made of the botanic gardens, veterinary hospital, forges and workshops. There is no entrance examination, but, unless under special circumstances, pupils must be at least 18 years of age, and they are expected to have received a good general education. In-students, about 85 of whom can be accommodated in the college-buildings, pay £135 per annum, or £45 per term, each having a separate "cubicle" and sharing a "study" with one companion, by an extra payment of ten guineas per annum a student may have a private room as sitting and bed-room combined, entirely to himself. Out-students pay £75 per annum, or £25 per term; they must be at least 21 years of age, and, except in special cases, must live in the town or neighbourhood in a house approved of by the principal. The ordinary college course extends over two years or six terms. Instruction is given both theoretically and practically; the former by lectures and laboratory practice, and the latter by demonstrations on the college-farm. The following is a synopsis of the order of studies :—Terms 1 and 2, Class I. : Agriculture (soils, manures, implements, labour, etc.), book-keeping, surveying, physics, geology or botany or zoology, mensuration, veterinary anatomy and physiology, drawing (plan).

Terms 3 and 4, Class II. : Agriculture (tillage, crops, etc.), chemistry (organic), book-keeping, surveying, physics, geology or botany or zoology, veterinary pathology, drawing (machinery).

Terms 5 and 6, Class III. : Agriculture (stock, dairy-farming, economics, etc.), chemistry (agricultural), book-keeping, levelling and engineering, physics, mechanics, geology or botany or zoology, veterinary therapeutics, obsetrics, etc., drawing (design).

Agricultural law in the winter session; building materials and construction in the spring session: and estate management in the summer session in each year The college farm is about 500 acres in extent, one-tenth only being in permanent grass. The students who desire it have opportunities of taking part in all kinds of work upon the farm, and, if qualified, can have special work allotted them, excellence in executing such work being rewarded. By arrangement with a prize dairy-farm at Bath, students of the college, upon their paying a fee of ten guineas, receive a fortnight's thorough instruction in cheese-making on the "cheddar system." Six scholarships, three of £25 and three of £10, are awarded annually. There is very little doubt that a young man can, if he likes, obtain at this college a thorough technical agricultural training, at the same time it must be noted that the fees it is necessary to charge, render its benefits unattainable to any but the wealthy.

The next in point of its period of formation is the Agricultural School at Aspatria, Cumberland, which was established in 1874, "for the purpose of affording pupils as much insight into agricultural practice as possible before they left school." The pupils first receive a good general education up to the age of 15, and then they proceed to the study of the various sciences bearing on farming. The elder youths work a portion of each day on one or other of the nine distinct farms connected with the school, under the personal supervision of the principal and practical farmers. The instruction qualifies for the Junior examination of the Royal Agricultural Society of England, the Bursary examination of the Highland and Agricultural Society of Scotland and for the certificate of the Science and Art Department. This institution has been largely assisted by the landowners and farmers of Cumberland and district. There are three terms in a year. In-pupils pay from 15 to 17 guineas per term, and out-pupils £3 to £5 per term.

The College of Agriculture, Downton. Wiltshire, was opened in 1880. Being a purely private undertaking, it has no public history; but there is ample to shew that it is, in a limited degree, an institution capable of giving useful agricultural education. Connected with it is a farm of 600 acres, which owing to the varied character of its soils and produce. is well adapted for educational purposes. Each in-student must have entered his 18th year, and find references as to his previous good conduct. Out-students must be over 21 years of age, and are required to be as regular in their attendance at lectures, classes, and examinations as in-students. The charge for the entire year is £129 for in-students, and £60 for out-students. Books, apparatus, instruments, and breakages being extra, in addition a fee of £1 is charged for the chemical laboratory. The course of instruction is as follows : Lectures, field classes, and practical work. The results of the teaching are tested by weekly examinations conducted on the farm, in the laboratory, and by means of printed papers. Each student is expected to keep a farm-journal. The subjects of instruction include agriculture, dairy and pastoral farming; estate management, land agency, and forestry; mensuration and land surveying; architectural drawing; book-keeping and commercial knowledge; physics and mechanics; chemistry; geology and mineraology; botany and vegetable physiology; zoology; veterinary surgery. The lectures and classes are so arranged that in a complete two-years' course a student has everything laid before him, to enable him to pass the senior examinations of the Royal Agricultural Society of England, of the Highland and Agricultural Society of Scotland, and of the Institute of Surveyors. The College and Training Farm at Hollesley Bay, Suffolk, is principally devoted to training youths for colonial life. An agricultural college has been established at Tamworth, and has in connection with it a 1,000-acre farm, corn mill, dairy, and workshops. This is a private speculation. These are the only really technical institu-tions for teaching agriculture to be found in England, that at Aspatria being the only one suitable for the rank and file of agriculturists. The words of Professor Wrightson, spoken in reference to Downton College, of which he is principal, are equally applicable to Cirencester: "The instruction is expensive to provide; the class of men who come must be housed and fed as gentlemen; hence the cost and the fees. It would answer no good purpose to lower the fees, as farmers, as a rule, would want them at least halving, and the college would thus suffer, and the landowners would not use it. I look upon this college as a school for landowners, land-agents, and colonists. I consider that, unless in the case of wealthy farmers, the agricultural education of farmers and farm-bailiffs falls outside our province."

There are, however, in existence other modes of imparting a certain amount of technical education on the scientific side, useful to those who can obtain practical knowledge on an ordinary farm.

The University of Oxford has long had an endowment for a chair of Rural Economy, but up till quite recently it was united with the chair of Botany. Since 1884, however, the two have been separate; and now the students have the privilege of attending a class on agriculture. The lectures are given in Magdalen College, and twelve is the minimum per annum, four in each term. As the course is so short, the subjects connected with agriculture must be taken very generally, or else only a very small department taken up. There is also a lectureship in agricultural science attached to Balliol College, intended especially for the Indian Civil Service probationers. Both these are open to students who take an interest in farming, and as these are mostly of the landlord class they can gain a certain amount of information which will be useful to them in after-life, though, of course, not by any means a thorough or exhaustive knowledge of the subject.

The Normal School of Science and the Royal School of Mines (amalgamated), London, some years ago established a

lectureship of agriculture, which later on was made a professor-
ship, where students could attend a course of about 40 lectures.
Three years are required to complete attendance at the
prescribed classes, and at the end of the time students can
graduate with the degree of "Associate in Agriculture." The
subjects taught are, drawing, biology, geology, mineralogy,
mechanics, and the principles of agricultural chemistry. The
fees for the first two years amount to about £75, and for the
remainder of the course they vary from £30 to £40.

The Institute of Agriculture, South Kensington, has been
established " for the purpose of advancing technical instruction
upon the various sections of agricultural practice as a preparation for learning the profession of farming, and especially for
bringing these advantages within the reach of any and every
person desiring to avoid much expense." Short courses of
lectures are instituted on specific subjects from time to time, and
gentlemen noted as authorities are asked to conduct a course on
their own special department. The full series amounts to about
280 lectures, at the rate of two daily by about 20 of the most
eminent authorities. The fees for the whole amount to only
£14. Also in connection with the Science and Art Department,
South Kensington, instruction in the principles of agriculture is
encouraged by the payment of fees to teachers in elementary
schools and science claases. The inducements offered to
scholars by this department are confined to prizes of books,
medals, and certificates, in addition to a certain number of
scholarships on condition that the scholar goes to some
recognised science school. Only three or four of these scholarships have hitherto been gained. In connection with this
department, evening classes are in many districts doing useful
work. Wherever a sufficient number of students can be got
together, a committee formed of men of local standing are at
liberty to appoint a certificated teacher to deliver one lecture a
week during the winter half of the year. The students are
expected to attend the examination in May, and the teacher is
paid by grant according to the success of the students. There
are some 400 of these classes in the United Kingdom, attended
by over 12,000 students. These classes are very suitable for
those who have left school without gaining any agricultural
knowledge.

In 1882 King's College, London, and the City of London
College each arranged for the delivery of a course of evening
lectures on the principles of agriculture. These have not up to
the present been very well attended. In both cases the lecturers
are agricultural chemists. In 1861 the Society of Arts included
agriculture in its scheme of annual examinations, and offered
two prizes of £5 and £3, in addition to certificates of proficiency. In each of the first three years only one candidate
entered for examination; in 1864 there were four; in 1865 there
was none, and so the attempt was abandoned. Although, as
regards agriculture, this society was not successful, credit is
nevertheless due to the council for having as long ago as 1861

attempted to encourage proficiency in agricultural knowledge.

The Royal Agricultural Society of England obtained its Royal charter of incorporation in 1840, in which, among the other objects for which this society was founded, the following is stated: "To take measures for the improvement of the education of those who depend upon the cultivation of the soil for their support." The early attempts of the council of this society to encourage agricultural education were neither particularly happy in their conception, nor fruitful in their results. In fact, the society has hitherto acted more in the capacity of an examining body than as a teaching institution. This statement must not be taken as implying want of appreciation for the good this society, no doubt, accomplishes by its yearly examinations and prizes; still at the same time many of its members, including the writer, consider that more of its funds and influence could beneficially be used in advancing agricultural technical education. The Yorkshire College, Leeds, has appointed two lecturers, one for practice and the other for scientific agriculture. Agricultural lecturers are also being appointed in a few counties in connection with the County Council grant.

Schools specially for imparting a thorough technical training in dairying, have been lately established in Cheshire and Derbyshire; and the British Dairy Farmers' Association are carrying out a scheme for founding a college of this kind near Aylesbury. There should be less difficulty in carrying on this branch of technical instruction, even without State aid, as about six or eight weeks is ample time to enable any young man or woman of ordinary intelligence to master all the details connected with the manufacture of dairy products, consequently the expense for each student need not be great. *Dairyschools.*

Considering that dairy-farming has of late years been the most profitable part of agriculture, and also with regard to the fact that the foreigner has beaten the home producer, owing to the superior quality of his produce; the failure of Mr. G. Allender in 1881 to found a British Dairy School cannot be understood. It was proposed to establish this school by a Limited Liability Company with a capital of £100,000 in £10 shares. Mr. Allender offered to subscribe for a thousand shares and to organise and manage the institution as managing director. Neither he nor the directors were to receive any remuneration for their services until after a dividend of 6 per cent. per annum should have been earned, and 1 per cent. placed to the reserve fund. No promotion money whatever was to be paid, and no unnecessary expenses were to be incurred, the object being to achieve, if possible, a public good. Attention was drawn in the prospectus to the fact that the value of our imports of dairy produce amounts to fifteen millions sterling per annum, to a large extent in consequence of the deficient technical education of our dairy farmers. However, the total amount of shares subscribed for (in addition to Mr. Allender's 1,000) was 1,695, of which only 765 are credited to landed proprietors, although nearly every landowner in the country was written to. The attempt was, therefore, abandoned. Mr. Allender's idea was to purchase an estate of between one and two thousand acres, and to furnish it with every well-tested appliance and improvement as to machinery and buildings; and to engage the services of the most efficient teachers obtainable here or on the continent, so as to ensure to the pupils the best technical information possible. The pupils were to be of two classes, as stated in the prospectus, thus :—" In the first and most important it is intended to give thorough practical information to sons of farmers and others, for a fee only to cover the cost of board and lodging—say £25 for the course of instruction—such course to extend over six months from the date of entry, and the pupils to be limited to such number as can be actually employed. The fee to be paid in advance, and the pupil to undertake to conform to necessary regulations and to perform all reasonable duties imposed upon him ; in fact to be exactly in the position of an apprentice. The other class to consist of such young men as prefer to remain for a longer period, and who will not be *obliged* to perform any actual work. These students will pay a fee of—say 50 guineas per annum—in return for the information afforded them, and make their own arrangements as to board, residence, etc."

It is, of course, impossible to say what measure of success this dairy school might have achieved if it had been founded. Possibly the scheme would have had greater chance of being successfully started had it been on a smaller scale. A moderate sized farm could easily have been turned into an institution for this purpose ; which could be extended without difficulty if it became necessary. There is no doubt that a number of small-sized school-farms scattered about the country would produce better results than one or two large ones on the scale proposed by Mr. Allender.

The University College of North Wales at Bangor has recently established an agricultural department, by means of a special fund. This fund for the session 1889-1890 amounted to £968, derived from two sources—the Board of Agriculture, which granted £400, and private subscriptions, amounting to £568. This fund is kept entirely distinct from the general college fund, which is increased by an annual Parliamentary grant of £4,000. The agricultural fund is administered by a special committee, composed chiefly of gentleman directly interested in agriculture, and nearly every agricultural district in North Wales is represented on it. This committee has full control of this fund, and directs in what manner it shall be expended. From the prospectus we learn that two things are aimed at in connection with this scheme of agricultural technical education. " First : to provide at the college as complete a training in agriculture and the sciences related to it, as can be obtained at any of the recognised agricultural colleges, and especially to provide such a training as would be suitable for bailiffs, land-agents, farmers, and young men who intend emigrating with a view to farming in the colonies. Secondly: to make the college a centre of agricultural education for North Wales, and to organise throughout the six Northern Counties of the Principality a system of instruction in connection with the college, to meet the special wants of each agricultural district, and supply a graduated system whereby pupils may pass from the school to the college." The Parliamentary Report on the Distribution of Grants to Agricultural Schools in Great Britain for the financial year 1889-90, contains the following reference to the agricultural work of this college :—

"The College possesses exceptional advantages for carrying on the work it has undertaken in this connection, for it is situated in one of the most thickly-populated districts in North Wales. Facilities are also claimed on its behalf for dealing with the bi-lingual difficulty, which is one of the great obstacles to the spread of agricultural education in the Principality; while the governing body of the college is thoroughly representative of every district in North Wales."

Arrangements have been made with local landowners and farmers by which farms in the neighbourhood of the college can be used by the lecturer in agriculture and the members of his class for the purpose of practical instruction. The out-college instruction in agriculture is being rapidly extended by the systematic establishment of fixed local schools or centres. There have already been three dairy schools started in connection with the college—at Sylfean, near Welshpool; at Lleweni Hall, near Denbigh; and at Bangor. Also local classes for the instruction of elementary schoolmasters in agriculture, chemistry, botany, etc., conducted by the college lecturers, extension lectures by members of the college staff; and field experiments supervised by college lecturers. Altogether there is no part of the United Kingdom where technical agricultural education is being carried on in a better manner than in North Wales.

The first attempt in connection with agricultural education Scotland. in the United Kingdom was made in Scotland by the founding in the year 1790 of a Chair of Agriculture in the University of Edinburgh. At first this chair had a stipend of £50 per annum attached to it, given by the late Sir William Pultenay. In 1868 the endowment of this chair was increased to £300 per annum, of which one-half is paid by the Government, and the other by the Highland and Agricultural Society of Scotland. The total emolument of the professor is now £370, and the fees payable by the students are four guineas per course of 100 lectures. The Highland Society endeavour to stimulate young farmers to attend the lectures by offering annually ten bursaries of £10 each. The number of students ranges between thirty and forty who are occasionally landowners and the sons of landowners; but the majority are sons of tenant-farmers. Generally the previous education of the students enables them to take full advantage of the instruction given; but there is no entrance examination, and no certificate of previous education is required. The course of lectures extends over one year, and is divided into two parts: one treating mainly of the principles of agriculture, the other of practice. Students passing the necessary examinations obtain a certificate which is recognised under the Privy Council regulations as qualifying the holder to teach agriculture and receive payment by results. In 1886 a degree of Bachelor of Science in Agriculture was established in connection with this college, on the same standing as the B.Sc. degree in other departments, and open to all those who have attended certain qualifying classes. The subjects prescribed are eight in number, which can be taken in two periods of four each. These are: agriculture (elementary), chemistry, botany, engineering, for the first part, with zoology as an alternative for botany; and agriculture (the whole subject), agricultural chemistry, geology, and veterinary science, for the final.

It is notable that this is the only degree in agriculture which is granted at an University in the United Kingdom, perhaps we may say in the world. And it is very appropriate that the University of the capital city of that portion of Britain, so long famous for good farming, should be the first to recognise in this way the importance of the subject, and accept it as of equal standing with every other profession taught within its walls. The cost of taking this degree will vary with the individual student, but may be kept down within a very moderate limit. In Scottish Universities all the students reside in their own private lodgings, and consequently the outlay for each winter session of six months will depend on the style of living adopted. The class fees for the degree are generally about three or four guineas each, with extra fees for matriculation examinations, etc., but altogether the cost of education need not exceed £40 per session, unless the student is personally extravagant. The whole course for the degree might be gone through by a farmer's son for less than £100.

Aberdeen University.

In the University of Aberdeen there is the Fardyne lectureship on agriculture; the chemical, botanical, and geological aspects of the matter being most largely dealt with. The lectures are not less than twelve in number, and are free and open to any one whether regular students or not.

Glasgow Technical College.

At the Glasgow Technical College there is a course of one hundred lectures on agriculture delivered during the winter six months. Agricultural students can attend the ordinary classes in botany, geology, engineering, etc., in common with those studying for other professions. The full course occupies three winters, and is intended to cover the ground taken up by the diploma of the Highland and Agricultural Society of Scotland, and qualifies students for the same, while the fees charged are very moderate, varying with the number of classes attended.

Ireland.

Thus far the position of agricultural technical education in Great Britain only has been dealt with. On turning to Ireland, we find this portion of the United Kingdom for a long time enjoyed exceptional advantages in this respect. In Ireland almost everything that has been attempted for the advancement of agricultural education has been done under Government supervision. This is a marked contrast to what has taken place in England ; at the same time it must in fairness be stated that the initiative in this direction was due to private enterprise.

The utility of affording useful instruction in agriculture to the small farmers and peasantry of Ireland has been urged for a very long period. Thoughtful Irishmen who knew the backward state of the Irish agriculture could not fail to perceive the great importance of instructing the people in this branch of industry. A select committee of the House of Commons in 1823 reported in favour of it, and again in 1830. However, as it has already been stated, the efforts of private individuals first gave effect to the recommendation of the committee. In 1826 some Ulster landowners formed themselves into a company, obtained land, erected house and buildings, and brought the Templemoyle Seminary into existence. The funds were raised by the issue of shares of £25 each, of which one hundred and sixteen were subscribed for, making a capital of £2,900. Unfortunately the shareholders made the mistake of attempting to found at the same time a school for youths intended for commercial and mercantile pursuits. This part of the scheme was soon abandoned, and after various negotiations the Grocers' Company made what appears to be a very generous agreement with the School Company, letting them 169 imperial acres of land for £30 a year. The Grocers' Company had previously subscribed £1,200 towards the expense of the school buildings, to which the shareholders, out of the subscribed capital, contributed a similar sum. The number of pupils provided for was one hundred. The maximum number, however, never appears to have exceeded seventy, the normal number being fifty. The deficiencies in the receipts necessitated appeals for subscriptions from time to time, and up to the middle of the year 1835 the total amount of money contributed to the institution, irrespective of pupils' fees, reached a total of £5,800. Orginally pupils could only be received on the nomination of a £25 shareholder, one pupil for each share ; but this rule was soon relaxed on account of the difficulty in obtaining pupils. In the report for 1835 it is remarked :—" The committee are aware that there are very few farmers who can afford to part with the work of one of their sons and pay a sum of £10 a year for his education ; but they would call on the landlords to assist them, with the confident hope that if this institution were released from debt, and the house full, they would soon be able to decrease the expense of education." For this fee of £10 the pupils received theoretical and practical instruction on the half-time principle, that is to say, one half their time was devolved to the actual work on the farm. For this work the pupils received board and lodging free. There is every evidence to show the valuable nature of the instruction given. In the course of the first 23 years of its existence this school educated as many as 800 pupils, and it must be regarded as being eminently successful in accomplishing the objects for which it was founded. That it was a financial failure is to be regretted, but cannot be a matter of surprise. The same result is an inevitable sequence to the establishment of an agricultural technical school suitable for ordinary farmers without a sufficient annual subsidy, because to attract pupils the fees charged must be almost nominal. These financial difficulties caused this school to be transferred to the Commissioners of Irish National Education in 1850, and in 1866 it ceased to exist.

In the year 1830 a commencement was made with the system of National Education in Ireland. The Education Commissioners, while acknowledging the utility of diffusing correct information on the subject of agriculture, at first confined their efforts to the establishment in 1838 of a school of agriculture at Glasnevin, near Dublin, where they took a moderate-sized farm and erected on it suitable house, buildings and offices. Two classes of persons received instruction, the object being to qualify them to teach agriculture in their several districts when they returned to their own schools. Second, a number of young men were received as agricultural boarders and trained as professional agriculturists, in the hope that they would afterwards be employed by the landlords of the country in instructing their tenants. The commissioners were engaged in pursuing their system of agricultural instruction, when at the end of 1843 the Royal Commission, known as the " Devon Commission," was appointed, and proceeded to make a searching enquiry into all matters relating to the agricultural interests of Ireland. Landed proprietors, land-agents, practical farmers, and others acquainted with the state of Ireland, were examined ; and it is worthy of note in connection with this subject, that in the report made by the commission there appears the statement that " no difference of opinion was found to exist amongst the witnesses as to the advantages to be derived from an extended establish-' ment of agricultural schools." The original idea of the Education Commissioners was to combine agricultural with literary instruction in as many of the rural national schools as possible. They were, however, induced to enlarge their plans. Applications were made from all parts of the country for aid towards establishing agricultural schools of a more comprehensive class than they at first contemplated. In their report for 1849 they say : " We have during the past year received a considerable number of new applications for grants towards the establishment of model agricultural schools. We have found it necessary to postpone our decision upon twenty of these applications." These were all ultimately granted. Altogether a large number of these agricultural schools were founded in various parts of the country. It was at this period that an adverse criticism from a select committee of the House of Commons on Miscellaneous Expenditure was communicated to the commissioners. The committee expressed grave doubts as to the policy (not wisdom, be it noted) of engrafting an agricultural department upon a national system of primary education. In reply, the commissioners took pains to assure the Government that they contemplated great advantages from the system of agricultural education which they had conceived for the country. The acclamation of Irish opinion in its favour caused the commissioners to extend it by means of provincial model farms, "for the instruction of young men as farmers, stewards, etc." From time to time they took leases of farms, twenty in all, in various parts of the country, and spent about £115,000 in erecting residences, farm-buildings, etc. Four of these farms were in operation in 1849, and in 1856 they were all in complete working order. Besides these model farms, which were the property of the commissioners, and entirely supported by them, numerous farm-schools were opened from year to year under private influence, which obtained aid from the commissioners towards their organisation and support. In the year 1850 the commissioners determined to offer substantial attractions to the guardians of poor-law unions throughout the country, to encourage agricultural education in workhouse schools. Wherever there was a farm of sufficient dimensions attached to a workhouse, the commissioners offered a gratuity to the teacher of the school for success in the management of the farm, and for giving efficient instruction in agricultural science to his pupils.

In 1857 an agitation, unfortunate for Ireland in its results, was commenced against State aid to agricultural technical education. The opponents of this system disputed the right of the state to train up farmers and stewards at the public cost ; and they had many influential, but short-sighted, supporters in the House of Commons. The Government from time to time was harassed in its defence of the system ; and in deference to the views of Parliament, successive Chief Secretaries were absolutely hostile to it. Mr. Cardwell especially directed his opposition to the countenance given by the commissioners to agricultural instruction in workhouse schools at the expense of the Imperial exchequer, as apart from the rates, and in 1862 successfully urged the commissioners to abandon it. Thus the workhouse experiment lasted only twelve years.

In 1870 the Royal Commission on Primary Education recommended "that the position of provincial and district agricultural schools in Ireland should be revised by the Commissioners of National Education, and that their number be reduced." In 1873 there were eighteen model farms in operation under the control of the commissioners. The number of national schools in which literary and agricultural education were combined, was at this date one hundred and fifteen, costing the state for the agricultural part of the education only £5 per annum.

Looking backwards, we at the present time can scarcely understand that any intelligent person should object to such an insignificant outlay for so useful an object as the diffusion among the Irish people of sound agricultural knowledge. But objections were made, and with such success that in 1880 only two institutions for affording technical instruction in agriculture remained, viz., the Albert Institution at Glasnevin and the Munster Farm at Cork. Both these still survive.

Some idea of the scope of the former may be gathered from the following extracts from its prospectus :—

" This institution is designed to supply instruction in the science and practice of agriculture to the sons of farmers, agricultural teachers, and others. An area of 6 a. o r. 7 p. (statute) is cultivated as a small spade-labour farm, with a view of exhibiting a proper system of cultivating the vast number of small farms in Ireland.

" An area of 22 a. 3 r. 7 p. has been set apart with a view of illustrating a system of farm management adapted to the circumstances of farmers whose holdings are large enough to give employment to one or two horses.

" The remaining portion of the farm, 162 acres, is cultivated as a large holding. The arrangements for affording the students as large an amount of information as possible upon every branch of the business of farming, including dairy-husbandry, the fattening of cattle, the breeding and rearing of different kinds of live stock, the various operations of field culture, and the permanent improvement of the soil, are such as to place within their reach an opportunity of becoming acquainted with the details of practical agriculture.

" In order that the students should have an opportunity of acquiring a knowledge of horticultural pursuits, about three statute acres are set apart and cultivated as a kitchen garden. There are also a small conservatory, peach-house, vinery, and fruit and flower gardens.

" The course of instruction includes a sound English education as well as agricultural science.

" Each of the lecturers of the institution delivers a course of lectures every session.

" In order that the students may become fully acquainted with improved practical husbandry, they are called upon to take part, for a limited time, in the performance of every farm operation, the feeding and management of stock, etc. They are also made practically acquainted with the uses of a large collection of improved farm implements and machines.

" There are two terms, of four months each, in the year.

" Three classes of students are admitted into the institution.

" I. Free resident students, who are boarded, lodged, and educated at the public expense, and who are admitted twice a year by competitive examination.

" Some respectable person must certify that the candidate's age is not under 17, that he possesses the necessary health and physical capacity for farming, and that he is of good moral character, and possesses industrial habits and tastes.

" II. Paying resident students, a limited number of whom are admitted on the following conditions : they must possess sufficient literary acquirements to profit by the lectures of the various professors. They must not be under 16 years of age.

" The fee for each session is £7 10s., which includes cost of instruction, board, lodging, etc. Paying students must take part in all farm operations with the free students. They take their meals at the same table with the free students, and receive the same treatment in all respects.

"The paying students whose conduct is satisfactory will be allowed to compete each half-year among themselves for a limited number of free places—one free place being reserved for every five paying students.

"III. Non-resident students. Young men who board and lodge at their own expense in the neighbourhood are permitted to partake of the advantage of the institution on the following terms :—

" 1. That while at the institution they shall be treated in every way like the resident class.

" 2. That they attend punctually all the lectures delivered at the institution.

" 3. That they be amenable to the rules and regulations.

" 4. That they pay in advance a fee of £2 a session."

The Munster Farm, near Cork, was turned into an agricultural and dairy school in 1880. According to the prospectus, there are two main departments of the institution, one the instruction and training of the sons of farmers and others in the best modes of developing the resources of the land, and the other the instruction of the daughters of farmers and of others in improved modes of dairy management. That the latter department has already produced practical results, is shewn by a marked improvement in the quality of the butter produced in the district. *Munster Farm.*

This institution receives a Government subsidy of £1,000 per annum.

In connection with this special branch of agriculture, a notice of Canon Bagot's " Travelling Educational Dairy" cannot be omitted. This, when closed for a journey, looks like a large furniture van. When required for work the wheels are taken off ; the upper halves of the sides are expanded upwards so as to form eaves to the roof, while the lower halves are let down to form a continuation of the floor ; the result being a covered shed with a boarded floor, measuring twelve feet by ten feet, and furnished with all the appliances necessary for dealing with the milk from a dairy of six cows, with hot and cold water, the latter laid on from an arrangement at the back. This educational dairy is the property of the Royal Agricultural Society of Ireland, and part of their scheme in adopting Canon Bagot's proposals was to let out the dairy and its staff, consisting of a labourer, a first-class dairymaid, and an assistant, to landowners or other persons willing to hire it, as well as to local agricultural societies and local committees constituted for the purpose. The charge for the use of the dairy is £7 per week, which includes the board and lodging of the staff. It is a great proof of the value of technical education, that wherever this travelling dairy has been located there has always followed a marked increase in the quality of the dairy products made in the neighbourhood. The cost of the dairy (exclusive of furniture), including hot and cold-water fittings, cistern, and boiler, was £115. *Canon Bagot's travelling dairy.*

We may now consider that we have reached the terminus of

our journey through the past and present of technical agricultural education, during which three prominent facts have been brought to our notice.

First : That facilities exist in Great Britain for acquiring a thorough technical knowledge of agriculture within the reach of those to whom expense is no object, while in Ireland this can be obtained free or at a merely nominal cost, but by a limited number only.

Second : That there is a comparative large amount of machinery for the delivery of lectures, which, however, does not appear to have produced much benefit, a result owing to the fact that farmers do not appreciate their advantages.

Third : That no system of agricultural technical education can be brought within the reach of the bulk of our agricultural population without either private subscriptions or state subsidy, or the two combined.

*In connection with State aid to agricultural education, it is interesting to note that on the continent this is the order of the day in most countries. For instance, in Germany the students in the higher agricultural schools (similar to our Cirencester and Downton) cost the state £100 per head per annum. The state subsidy to the intermediate agricultural schools amounts to £155,000 per annum, and in the lower to £17,000 per annum. In France, too, state aid for this purpose is on a similar scale. It is beyond the scope of this essay to give any detailed information as to the various ways in which continental countries give encouragement and help to technical agricultural education ; but the writer would recommend any one interested in this part of the question to make themselves acquainted with the facts under this head contained in the second report of the Royal Commission on Technical Instruction.

Future.

We are glad to see that Mr. Chaplin, in his capacity of President of the Board of Agriculture, realises the importance of more attention being given to this question in the future than has been the case in the past. In a letter addressed last year to the Chancellor of the University of Cambridge, Mr. Chaplin appears to think that this University might do something towards promoting the spread of scientific instruction among our agricultural population. Mr. Chaplin does not say whether he is referring to schools or colleges, or to any organised system of lectures ; but he points out that " an agricultural teacher, to be successful, must be master not only of the ordinary methods of agricultural practice, but must, in addition, have intimate acquaintance with those sciences, such as chemistry, botany, geology, biology, etc., on the principles of which a correct and intelligent practice of agriculture mainly rests." The answer to the question as to what extent it may be possible for the University of Cambridge to contribute to the diffusion of agricultural knowledge among the rising generation of farmers depends a great deal on the manner in which it is proposed that

* See Appendix C.

this instruction shall be given. But there is this that is certain that some of the funds at the command of our Universities might well be spent in assisting some general scheme of agricultural technical instruction. On November 3rd, 1891, the Agricultural Education Syndicate of the University of Cambridge recommended the appointment of a lecturer in agricultural science, who shall also be director of agricultural studies, at an annual stipend of £500, and a lecturer in agricultural science at an annual stipend of £300, to continue so long as the payment of £400 annually by the Cambridgeshire County Council is continued. The syndicate hopes for contributions from other neighbouring counties, and that the Government may be induced to make a grant in aid.

It has already been pointed out that in the term agriculturist is included the landowner, the tenant-farmer or occupying owner, and the labourer. Although the main interests of these three classes are identical, yet the character of the technical education which should be given to each, must obviously be very dissimilar.

Landlords may be said to be already provided for (at least such of them who care that their sons shall receive some agricultural knowledge) at Cirencester, Downton, Aspatria, etc.; institutions which, if the demand arose, could be easily increased. Apparently, however, very few of this class care a great deal about technical instruction, yet there is very little doubt that the scions of the landed proprietors would be all the better for a few years spent at colleges of this kind, rather than wasting the earlier parts of their lives, as is too often the case, in loafing about at clubs, racecourses, and other ways which the aristocratic unemployed have of killing time. Were landowners, as a whole, better educated in agricultural matters, we should not find them compelling their tenants to farm under such antiquated, restrictive covenants; they would be alive to the necessity of providing better homesteads where the manufacture of meat could be carried on at minimum of expense; covered yards would be the rule, not, as they are at present, the exception, and they would be more likely to take greater personal interest in their estates.

In any scheme for increasing, in the immediate future, the quantity and improving the quality of technical instruction suitable for the farmer, be he large or small holder, the wants of two sets of individuals must be considered. One, farmers' sons or other lads intended to be brought up as farmers; the other, those who are actually engaged in trying to gain a livelihood by cultivating the soil, but who have not received any special education for their profession. For the latter, lectures and evening classes appear to be the only means of supplying their need. The University extension system could well be utilised for this purpose. Hitherto agriculture has been conspicuous by its absence in the courses of lectures delivered under this scheme. Where University extension centres have been established, the cost to the students averages about one shilling per

head for each lecture, generally, however, it has been necessary
to obtain private subscriptions to assist in defraying expenses.
This, coupled with the Science and Art Department,* furnish all
the machinery required for a complete system of agricultural
lectures throughout the country, in sufficiently numerous locali-
ties to be within the reach of every farmer. In connection with
the expenses attendant on a complete scheme of this kind, the
powers lately conferred on County Councils are very important.

The general effect of the "Technical Instruction Act" is to
empower any County Council to apply the proceeds of a penny
rate, and also the sum falling to its share from the £743,000
provided by the new beer and spirit dealers (after the Police
Superannuation Fund has been deducted), to supply such
technical instruction, whether industrial, commercial, or agricultural, as is most required for the district. On mixed councils
of large counties, which contain at the same time large manu-
facturing towns and extensive agricultural districts, "technical
education" means to the majority of the members instruction
for future artisans; there is, therefore, a danger that the
advantages which are within the grasp of the agricultural
community being allowed to pass by. It must be remembered
that this act is purely permissive, so that without pressure being
put upon their members by agricultural constituents, it is very
likely that the need and necessity for such instruction in rural
districts will be overlooked altogether.†

The portion which agriculture can justly claim of the sum
raised under this Technical Instruction Act will doubtless be
found insufficient for providing agricultural lectures in every
village; at the same time if the fund is properly managed, it will
go a considerable way in this direction.

For lads, schools similar to that at Aspatria would be
excellent. Students from this college have been more successful
than those from any similar institution. Although the technical
education at this institution commences at the age of 15, which
is two or three years earlier than it is at any other agricultural
school in the country, still the writer is of the opinion that this
instruction could well begin at the age of 13. A boy of this
latter age could easily commence to learn something about the
profession he is intended to occupy, and there are many light
occupations on a farm or garden which he could easily perform.
But apart from any considerations of his future occupation, a few
hours daily spent in some physical employment would increase
his mental capabilities, so that he would be able to accomplish
more of the ordinary school work with greater ease. Although
this school at Aspatria is successfully carried on without State
aid, it must be repeated that this is only achieved by means of
private subscriptions.

‡There is no doubt that it would be an immense boon to
agriculture if one school of this kind were established in every
county, and there is certainly no reason why this should not be

* See Appendix D.   † See Appendix A.   ‡ See Appendix B.

the case. Agriculture itself with a moderate amount of State aid could easily establish and carry on technical schools sufficient for its requirements.

The farm, which should be about 300 acres in extent, attached to the school should, of course, pay its way. This should be the first principle of the whole matter that farming should be taught *on a farm that pays*. The expenses of the school, over those of an ordinary farm, would be the cost of the teaching staff, accommodation for pupils, a prize fund and books, etc. In connection with finding the money necessary for this. it may be asked whether a portion of that now spent in agricultural shows could not be diverted to technical education ? Agricultural shows may have some beneficial effect on agriculture, but it is questionable if the good results are in any degree proportionate to the cost. Take for instance, the Royal Agricultural Society of England which spends about £20,000 per annum in connection with its annual show, couple to this the amount expended by all the other agricultural societies, and we have a sum of money which if only half of it were spent in the cause of technical education, the results would be ten times greater.

It may, however, be argued that if you wish farmers' sons to learn how to farm, they can do so at home. The question is, can they learn how to make farming pay at home ? Is it not a fact that, in the majority of cases, farmers themselves know very little of the principles of agriculture ; therefore they cannot teach their sons the how, the why, and the when, of even a portion of the operations performed on a farm. The necessity for school-farms is, therefore, apparent, so that the theory and practice of agriculture, in its most advanced state, can be taught simultaneously.

It may be mentioned, in passing, that in France school-farms known as *Ecoles pratique d'Agriculture*, exist very largely. The pupils enter at the age of 15, and pay a fee of £16 per annum which includes board, lodging, and washing. The pupils have to assist in the farming operations. The cost to the state being about £31 per pupil per annum.

To the agricultural labourer the question of technical education is important from a very matter-of-fact standpoint. A lad taught, as he might be, a thousand things that would be of help on a farm, could not fail to be worth more money to his employer than the uninformed rustic, who through ignorance may possibly spoil the work of a year in the course of a day.

The Agricultural Labourer.

The statement that education is spoiling all classes of servants is true only in the sense that the education given at the elementary schools, more especially in the rural districts, is not of the most useful kind ; but no one can argue that intelligent labour is inferior to unintelligent.

The agriculture of a country must ever be largely affected by the condition and character of its peasantry by whom its labours are performed.

An acute observer has pointed out that a poor style of

farming and low wages—that good farming and comparatively high wages, usually go together. The eminence of the agriculture of Scotland is due, in a large measure, to the moral worth and intelligence of her peasantry.

Our elementary education system is not doing anything like the good that was expected of it.

The children of the agricultural labourers are entering life without that technical knowledge which is fast becoming an absolute necessity for success in any calling. The need for this is the more urgent now that labourers can obtain allotments. There is no doubt that in the Reformatory schools conducted on the "farm system" we have precisely that control over a child's education at a critical period of his life, viz., from 12 to 14 years of age, which is lost in elementary schools conducted as at present. No doubt the education given at our Primary schools fits a boy to be a clerk, but is the country in need of clerks? Do we not require more productive labour? The old notion that it is gentlemanly to be a clerk, and degrading to be a workman, is dying slowly. Mr. James Scotson, headmaster of one of the largest schools in the north of England, the other day remarked: "I have had considerable experience both of parents and scholars, and I find a great many parents are beginning to see that competition among clerks is much keener than in any other field of work. If ever the dignity of hand labour is to be thoroughly recognised, we must begin at our elementary schools, by showing that it is as honourable and as useful to use a tool as a pen, in connection with our elementary education." The country greatly needs a better class of workmen, who will strive to turn out the best possible quality of work. A comparison was made at a public meeting at Manchester between the two characters by Mr. F. Smith, an architect. "If," said Mr. Smith, "I met two young men in the street, one with a tremendous thick stick and a cigarette, and the other with a bag of tools in his hand; if I were to take off my hat to either of them, it would be to the man with the tools in his hand."

On all sides educationalists are calling attention to the value of physical training as a means of increasing the mental capacity. Agricultural work is specially adapted for the lads of our schools; carried out judiciously, it should result in the combination of a sound mind in a sound body. Physical exercise in the open air will develop the one; while the endless variety of occupation and the subjects of interest connected with it, wisely handled as a means of educating, will seldom fail in producing the other. For a boy who will have to earn his living by manual labour, the lessons to be learned on a farm or garden will be always useful; to such as emigrate they are invaluable. A well-managed piece of land is one of the best means of affording technical instruction, and, when properly utilised with that end in view, would certainly place the boys trained on it in a position far in advance of labourers who had had no similar advantages. Practical work by no means interrupts or even retards elementary education. Where this has been tried in industrial schools, masters have reported that no difficulty is found in moving boys of ordinary capacity one standard a year, even while devoting a large proportion of their time to specific subjects directly bearing on their practical training. In dull boys this practical training is the most useful part of their education; they do, under this system, learn to use the faculties they are endowed with, instead of spending their time in a vain endeavour to develop powers they do not possess. There is no question as to the adaptability of farm labour as a means of influencing the mind in the right direction. Among those who have given any of their time to impartially studying the question, there is a great preponderance of opinion in favour of combining agricultural training with elementary education. The experience gained in the reformatories and the results attained should help much to guide us in formulating and adopting any system of technical education for the children of the working classes. In these industrial schools the least promising portion of the community are dealt with, but the results are extraordinary.

Of course it does not necessarily follow that book-work in rural elementary schools should be restricted to three hours a day, as it is in reformatories, but at the same time there is no doubt that the principle of the half-time system, which has been advocated by many for more than a quarter of a century, is the right one, and its success with comparatively young children in the industrial schools is very striking. By the adoption of such a system in rural elementary schools, boys would learn to use their heads and hands in combination, and thus discharge their duties intelligently, instead of mechanically.

There is no reason why the system of general instruction pursued should not, from the earliest period of a child's education, lead up to technical knowledge. For this purpose the reading books used in primary schools, more especially those in rural districts, should contain pieces on subjects likely to be of use to the scholar in after-life. At present these books contain, for the most part, mythical, fabulous, and poetical matter, anything calculated to increase the child's power of, in after-years, successfully meeting the competition of foreign labour, is conspicuous by its absence. Surely it is a fact that from the agricultural labourer upwards to the largest farmer, all must either turn out the articles they produce at a cost enabling them to sell as cheaply as the foreigner, or they must cease to be producers; it is not to be expected that the general body of consumers will pay more for English produce than for the same quality of foreign. There is also this other fact which is equally true, that this country could produce a far greater amount of food than it is now doing, and many times the present number of people could find *profitable* employment on the land. But to this end, more especially if small holdings are to increase, the cultivators of the soil must start with some knowledge of the art of husbandry. To commence the inculcation of this technical knowledge as early as possible, why should not reading books in elementary schools contain pieces in simple language on such subjects as animal and insect life on the farm; vegetation, distinguishing between weeds and cultivated plants, the cause of the seasons, etc. These and others could well be found in reading books for Standards II. and III. For Standard IV. the books should contain information on these subjects of a more advanced kind and, in addition, elementary botany; the action of the different classes of soils on vegetation; how soils are formed, etc. At this period (Standard IV.) should also be commenced the imparting of the practical knowledge of cultivating the soil. For this purpose every elementary school in rural districts should have a piece of land attached to it, varying in size according to the number of boys attending. A certain number of hours per week should be spent in this school garden under the supervision of a properly-qualified instructor. There is no reason why all elementary school masters should not be possessed of the necessary knowledge to instruct in the principles of cultivating the soil and the use of tools. It is obvious that a greater portion of time should be spent in this way in summer than in winter. During the latter period or at any other time when it is not possible to work in the open, instruction should be given in some other useful branch, such as carpentry, a knowledge of which is required as much on a farm as anywhere.

It may be mentioned that in France there are thirty thousand primary schools with pieces of land attached to them for educational purposes, and in these, practical instruction in cultivating the soil commences at nine years of age; also all teachers have to pass in agriculture previous to appointment.

After a boy has passed the fourth standard, or has received instruction in the school garden for twelve months, he should be allowed to obtain employment on the half-time system. As a matter of convenience, it would be better for one portion of the "half-timers" to attend school in the morning, and the other in the afternoon. There is very little doubt that if this system were carried out and parents found that their boys were gaining useful, instead of useless, knowledge, they would not take them from school at the earliest possible moment, as is now generally the case. There are those who argue that the age and standard at which children should be allowed to leave school should be raised, but this is decidedly impracticable and also unfair, as it would deprive parents of the six and more shillings a week, which an intelligent lad can earn in rural districts directly he leaves school at the age of 14. Lads over this age could obtain further technical education at evening classes. There is no doubt that in connection with the latter the magic-lantern will come more and more into use to illustrate lectures. The attendance at these evening classes, properly organised and under the direction of a duly-qualified instructor, should be made compulsory, say up to the age of 16 years, after which those who wish should be allowed to continue their attendance. The question of fees will, of course, have to be considered. As elementary education is now free, and the evening classes would come at a time when the lad would be earning wages, a small fee could be paid easily, and could be demanded with justice. Further financial help might be obtained by diverting some of the money squandered by school-boards in the present useless system of cramming.

All evidence points to the fact that a complete system of technical instruction should provide a fitting preparatory training to the different classes of persons engaged in productive industry. The programme of elementary, as well as of secondary and higher education must be organised with reference to their different special requirements. If the demand for technical instruction is to be fully satisfied, a great part of our existing system of education must be reconstructed, and the training in the schools be made a more fitting preparation for the work of life than it is at present.

APPENDIX A.

The following will show how the new fund under the Local
Taxation Act, 1890, has been appropriated.—

County Council.	Apportionment of the £743,200.	Amount granted for one year to Agricultural Technical Education.
	£	£
Bedford - - - -	4,343	1,650
Berks - - - - -	6,478	1,730
Buckingham - - -	5,372	nil.
Cambridge - - -	3,379	2,240
Cheshire - - - -	17,576	4,050
Cornwall - - - -	6,282	750
Cumberland - - -	5,786	900
Derby - - - - -	10,924	no return.
Devon - - - - -	14,547	nil.
Dorset - - - - -	6,195	5,300
Durham - - - -	16,399	no return.
Essex - - - - -	21,225	(about) 2,000
Gloucester (County)	17,456	700
Hereford - - - -	4,877	375
Hertford - - - -	6,492	340
Huntingdon - - -	1,942	nil.
Kent - - - - -	22,962	(about) 5,000
Lancashire - - -	87,977	9,250
Leicester - - - -	7,612	300
Lincoln - - - -	12,403	nil.
London - - - -	163,192	nil. (all to rates)
Middlesex - - - -	20,889	nil. (all to rates)
Norfolk - - - -	10,759	1,500
Northampton (Co.) -	6,641	No fixed sum, a few lectures on dairying only having been
Northumberland - -	10,972	no return. [given.
Nottingham (Co.) -	9,356	435
Oxford (Co.) - - -	5,710	1,314
Rutland - - - -	706	400
Salop - - - - -	6,543	1,300
Somerset - - - -	13,994	1,000
Stafford - - - -	20,196	Two lecturers appointed, but no special sum set apart.
Suffolk (East) - -	4,772	250
„ (West) - -	2,142	890
Surrey - - - - -	18,304	3,150
Sussex (East) - -	10,949	Under consideration.
„ (West) - -	3,455	„ „
Warwick - - - -	19,142	nil.
Westmoreland - -	1,638	750
Wilts - - - - -	9,253	3,000
Worcester - - - -	9,696	1,000
York (East) - - -	7,306	1,300
„ (North) - - -	7,326	nil.
„ (West) - - -	47,045	1,500

In addition to the foregoing information in the third column, which is the result of direct inquiries made to the various clerks of County Councils, the writer has also been favoured with the following further details of the manner in which it is intended to spend the sum granted in aid of agricultural education.

In Cumberland the amount is divided as follows: £300 for salary and expenses of a peripatetic lecturer in agriculture; £400, salaries and expenses of migratory dairy school; £200, expenses of two classes in agricultural science for elementary school teachers.

In Dorsetshire the County Council has a very elaborate scheme before them, which reflects great credit on the committee and organising secretary who are responsible for it. This was presented to the County Council early last November. The proposed annual grants in aid of agricultural technical education are: University extension lectures, £1,000; dairy instruction, £500; plant for dairy school, £320; scholarships at cheese schools, £378; to secondary schools for providing plant and apparatus, establishing or perfecting agricultural sides to such schools, £1,500; scholarships to enable boys from elementary schools to proceed to secondary schools as above, £856; and £800 for manual instruction which includes agricultural engineering, sheep-shearing, hedging and ditching, working of allotments and cottage gardens, budding and pruning, etc. In this scheme the grant for manual instruction is too small; it would be an improvement if £500 were taken from the extension lectures and the amount added to the manual instruction fund.

The Essex County Council have a scheme under consideration, the principal points of which are that £500 be given for instruction in manual work and domestic economy; £1,000 for lectures on farriery and mechanics and towards establishing a dairy school; and £50 for lectures on bee-keeping.

In Hertfordshire the small sum has been chiefly expended in holding dairy classes for about ten days at the chief towns, but this county not being a dairy one they were only sparsely attended. The question of further grants for agriculture is now being considered.

The Lancashire County Council has granted the sum of £1,250 for migratory dairy schools; £2,000 for lectures and scholarships; and £6,000 has been allotted to the rural districts, the greater portion of which will be expended in carrying on classes in subjects relating to agriculture.

In Leicestershire the grant has been devoted to giving lectures on dairying.

In Nottinghamshire the County Council has granted £75 for expenses connected with lectures on practical agriculture; £160 for lectures on agricultural chemistry; £200, elementary schoolmasters' scholarships in agriculture.

The County Council of Rutland has devoted £130 for combined lessons in butter-making and cookery, and has set apart £270 more for purposes in this direction.

In Somersetshire £1,000 has been given for dairy instruction, and the additional sum to be given to general agricultural education depends upon the reports of various local sub-committees.

The East Suffolk authorities have done nothing at present, but give ploughing prizes.

Surrey has granted £500 for dairy-work; £2,500 for agricultural sciences; £150, agricultural manual work; it is also intended to give certain scholarships, but these are at present undetermined.

The County Council of Worcestershire has given their grant entirely to the Worcestershire Chamber of Agriculture, presumably to spend in aid of agricultural education in what manner they think fit.

The Wiltshire Council has granted £700 for dairy schools, the remainder to lectures on various agricultural subjects.

In Yorkshire (W.R.) the grant is divided into £156 for dairy schools; £844, extension lectures; £90, scholarships; £250 railway fares of teachers; £100, experimental plots.

Also several County Councils have granted sums for cookery, laundry work, and domestic economy in rural districts, notably, Dorsetshire £900, Wiltshire £700, Lancashire £1000, Rutland £130, combined with butter-making. This is, of course, for the female portion of the rural population, and although not strictly agricultural education, is worthy of consideration in connection with it, especially cookery, as there is very little doubt that in many agricultural labourers' homes there is as much food wasted by bad cooking as the majority of the peasants on the continent of Europe have to live upon.

Those County Councils which have adopted the most complete schemes, appear to have appointed organising secretaries with a special committee composed of C.C.'s, who have again appointed local sub-committees, the latter being composed of others besides the C.C.'s for the different districts.

The great fault of nearly all County Councils is that sufficient attention is not given to the importance of creating better workmen by means of manual instruction, so as to make it worth the while of occupiers of land to pay a sufficiently high wage to counteract the attractions of town life.

The necessity of founding agricultural school-farms appears to have been entirely overlooked. The want of these is one of the great drawbacks to rural life.

However, upon the whole, we are progressing in the right direction; perfection will not, of course, be attained to at once, mistakes and a more or less waste of money have already taken place, but so long as county councillors keep the idea of securing the maximum amount of technical instruction, in its truest sense, to the rural population, with a minimum of unnecessary expense, before them, they will, no doubt, in time arrive at a satisfactory solution of the problem; bearing in mind that no one is entitled to give technical instruction in agriculture unless he combines scientific knowledge and a thoroughly practical acquaintance with the mechanical—whether performed by man or machine—work on a farm.

It will be noted that several County Councils have given nothing for the purpose of agricultural education, it is to be hoped that a different state of things will exist after next March.

In those counties which have "no return" against them, the various clerks have refused to grant the information asked for by means of reply postcards.

APPENDIX B.

Approximate cost of establishing and carrying on a school-farm suitable for supplying technical instruction in agriculture to farmers' sons and others—

CAPITAL ACCOUNT.

	£
(1) Cost of purchasing a 200-acre farm, with house and buildings on it, at £20 per acre - - - - - - -	4,000
Cost of providing acccommodation for masters and, say, 40 resident pupils, also lecture room, workshops, etc.	2,000
Furniture, fittings, etc. - - - - - - - - - - - -	700
Cost of stocking the farm, implements, live stock, and working capital, at £10 per acre - - - - - - - -	2,000
(say £9,000)	8,700

(1) A great many farms have been sold during the past few years for half this sum per acre.

Yearly expenses, over and above those necessary for working the land as an ordinary farm—

Salaries of teaching staff : £

		£
Principal - - - -	200, with house and board.	
Assistant - - - -	70, ,, board and lodging.	
,, - - - -	50, ,, ,, ,, ,,	
Dairy instructress -	40, ,, ,, ,, ,,	
Assistant ,, -	25, ,, ,, ,, ,,	
Workshop instructor	80, find own board, etc.	
	465	465
Board and washing for, say, 40 persons at 15s. per week each - - - - - - - - - - - - -		1,560
Interest on extra capital, £3,000, at 5 per cent. - -		150
Prizes and incidentals - - - - - - - - - -		100
		2,275
Deduct for saving in board, etc., during holidays - -		150
Yearly expenses - - - - - -		£2,125

Probable yearly income——   £
There should be no difficulty in obtaining 10 per cent.
 on the capital cost of purchasing and stocking the
 farm, which on £6,000 would give - - - -   600
Fees from pupils .   £
 30 residents, at an average of £25 each   750
 10 out-pupils, at £10 - - - - - - 100   850

     £   £1,450
Expenditure - -   2,125
Income - - -   1,450

    £675—deficit to be supplied by State aid
and subscriptions.

Perhaps State aid would be best given by finding the capital in the first instance, leaving any deficiency in income to be supplied by agriculture itself. This latter could easily be accomplished by diverting some of the money spent in agricultural shows. The establishment should be under the control of a board consisting of two persons appointed by the Board of Agriculture, two by the Royal Agricultural Society of England* (provided this society assists financially), and two to be elected by the votes of owners or occupiers of agricultural land exceeding half-an-acre in extent, owned or occupied in the county or district where the school-farm is established.

The principal should be over 35 years of age, have spent three years at an approved agricultural college, hold the diploma of Royal Agricultural Society of England or of the Highland and Agricultural Society of Scotland, and the B.Sc. in Agriculture, Edinburgh. He should also have been actually engaged for at least five years in practical farming, since leaving college. He will have entire control of the management of the farm, buying and selling live stock, etc., in addition to the school. Of course under him there will be a good bailiff.

The first assistant-master should be fully qualified to instruct in botany, geology, general and agricultural chemistry, and animal physiology. The second should be capable of giving instruction in agriculture, horticulture, forestry, land-surveying, levelling, mensuration, agricultural mechanics, steam-engine management, etc.

The dairy instructress should hold a first-class certificate, and should have been engaged in the management of a dairy for at least three years.

The workshop instructor should be capable of performing all kinds of forge-work, including tempering, taking to pieces, and refitting machinery, etc. An efficient carpenter and wheeler should be employed to give instruction in his department. Both these persons could fill up their time by repairing machinery, etc., for neighbouring farmers, and would thereby add to the revenue of the establishment.

* Or Scotland or Ireland, as the case may be.

Regulations in connection with resident pupils—

No pupil to be admitted under the age of 14.

Pupils under 16 years of age to pay a fee of £25 per annum. Over 16 years of age, to be divided into two classes, as regards fees only.

I. Those willing to occupy the position of apprentices taking part in all the work of the farm, to pay a fee of £10 per annum.

II. Those who will not be compelled to perform any actual work, but who will be under the same regulations in other respects, to pay a fee of £50 per annum.

Out-pupils—

Over 14 and under 16 years of age, £5 per annum. Over 16 years, £15 per annum.

Credit has not been taken in the above figures for the value of work performed by apprentices.

Evening classes could be arranged for persons over 25 years of age.

There are many other minor details necessary to be taken into account in connection with a scheme of this kind, these, however, need not be set forth here.

APPENDIX C.

State aid to agricultural technical education, and fees paid by pupils in Continental schools—

	Fees, per annum. £	Cost to State per pupil per annum. £
GERMANY :		
High School of Agriculture, Berlin	7	100
Other higher agricultural schools, 12 in all	7 to 9	65 to 75
Intermediate and lower	5	10 to 15
Dairy schools	7 to 18	10
FRANCE :		
Higher { Institut Agronomique	12	105
Grignon	48 (including board, etc.)	80
Grand Jouan	40 ,, ,,	92
Montpellier	40 ,, ,,	72
Intermediate schools	2 to 15	30
Lower	Fees nominal, pupils work on farm	10 and salaries of staff
Special schools, forestry, etc.	20 to 30 (including board)	Total cost to State, £15,000 per annum
DENMARK :		
Agricultural and Veterinary College	Free	30
Farm-schools	35s. to 40s. per week, including board, etc., pupils working on farm	1 to 2

BELGIUM :   £
Institut Agricole -   -   -   -   28 (including board) 60
Horticultural School-   -   -   4 to 8   55
     State aid is also given to experimental stations.

     NETHERLANDS :
Rykslandbonw School -   -   6   50
State Agricultural School -   40 (includ-   State supplies
                                ing board)   entire defficiency
Lower schools   -   -   -   -   20   95

APPENDIX D.

Districts where agricultural science classes were held, with number of students, for the year ending 1890, in connection with the Science and Art Department, South Kensington—

	No. of Students.		No. of Students.
Bedfordshire.		**Durham.**	
Bedford	6	Blaydon-on-Tyne	15
Potton	8	Crook	10
Berkshire.		South Shields..	27
Reading	6	Stockton-on-Tees	30
Buckinghamshire.		Sunderland	40
Nil.		**Essex.**	
Cambridgeshire.		Canning Town	25
Cambridge	10	Chelmsford	70
Willingham	15	Stratford..	16
Cheshire.		**Gloucestershire.**	
Crewe	20	Cheltenham	20
Cornwall.		Gloucester	30
Camborne	12	Sydney	20
Helston	10	**Hampshire.**	
Liskeard	20	Purbrook	12
Cumberland.		**Herefordshire.**	
Carlisle	12	*Nil.*	
Derbyshire.		**Hertfordshire.**	
Bradwell..	20	Barnet	25
Chesterfield	12	**Huntingdonshire.**	
Clay Cross	20	*Nil.*	
Lea	12	**Kent.**	
New Mills	20	Canterbury	10
Devonshire.		Chatham..	10
Barnstaple	8	Deptford..	50
Crediton	10	Rochester	6
Exeter	6	Swanley	40
Plymouth	17	Sydenham	20
Dorsetshire.		Tunbridge Wells	10
Bridport	22	Woolwich	20
Charminster	14	Yalding	20
Gillingham	4		
Sturminster	8		

Lancashire.	No. of Students.		Nottinghamshire.	No. of Students.
Cranshawbooth	30		Eastwood	40
Faringdon	10		Oxfordshire.	
Heaton Chapel	50		Witney	10
Lytham	10		Salop.	
Middleton	25		Cleobury Mortimer	20
New Hay	8		Madeley	12
Preston	20		Somersetshire.	
Rochdale	8		Radstock	6
Ulverston	12		Staffordshire.	
Walsden	12		Wolverhampton	8
Leicestershire.			Suffolk.	
Lullerworth	20		Gorleston	12
Lincolnshire.			Surrey.	
Grantham	15		Croydon	20
Lincoln	12		Newington	30
Moulton	6		Peckham	30
Stamford	8		Sussex.	
Middlesex.			Brighton	32
Chancery Lane	50		Warwickshire.	
Hackney	40		*Nil.*	
Hammersmith	12		Westmoreland.	
London Wall	50		*Nil.*	
Mile End	10		Wiltshire.	
Moorfields	25		Salisbury	2
Notting Hill	6		Worcestershire.	
Paddington	8		Dudley	6
Regent Street	12		Yorkshire.	
Strand	60		Bradford	30
Tottenham	15		Calverley	6
Waltham Green	36		Easingwold	7
Monmouthshire.			Elland	6
Abertilly	10		Great Ayton	35
Tredegar	15		Hebden Bridge	8
Norfolk.			Hull	78
Necton	12		Ilkley	20
Norwich	32		Middlesborough	24
Northamptonshire.			North Cowton	20
Brackley	18		Sheffield	60
Kettering	10		Skelton	7
Oundle	24		Skipton	15
Northumberland.				
Newcastle-on-Tyne	5			

North Wales—8 classes, with 92 students.
South Wales—11 „ „ 178 „
Scotland —84 „ „ 1733 „
Ireland —73 „ „ 5669 „

Total number of students of agriculture in the United Kingdom in connection with the Science and Art Department—9750.

APPENDIX E.

SPEECH BY THE RIGHT HON. W. E. GLADSTONE, M.P.

The following is a portion of a speech made by Mr. Gladstone at Chester on the occasion of the show of the Royal Agricultural Society of England in the year 1858, which is worthy of notice in connection with the necessity for education if the best results are to be obtained from farming ; and also with regard to the importance of agriculture as an industry. It is taken from the *Mark Lane Express*, of November 24th, 1890, the Editor of which, before republishing, sent a proof to Mr. Gladstone, who, in returning it, sent this in reply:—

"Dear Sir,

"It cannot be otherwise than a pleasure to me if you deem a speech of mine touching upon agriculture, and now thirty-two years old, worthy of reproduction in your columns.

"I had then and have now a high sense both of the difficulty and the dignity of the agricultural calling. I rejoice that the severe pressure of late years has neither destroyed, nor, as far as I can see, impaired the union between occupiers and landlords ; and I rejoice to think that the cultivators of the soil are improving their position as compared with that of a short time back. We have all, in all callings, much to learn from misfortune, which teaches us to husband, not to waste, our resources, and to be ever on the watch for suggestions and opportunities of improvement.

I have the honour to be, Sir,

Your faithful servant,

Hawarden,   W. E. GLADSTONE.

November 17th, 1890."

THE SPEECH.

"I have the honour to propose to you as a toast, a society founded under the highest auspices. . . . . This society is so founded, and so combines the universal suffrages of the country, because it is directed to such a purpose as that of promoting the most essential and the most venerable among all the arts that furnish material for the industry of man. Whatever else may come, and whatever else may go, this at least we know that no vicissitude of time or change can move agriculture from the position it has ever held, and ever must hold until the last day in the crack of doom itself. . . . . We must not suppose that because it is an ancient art, and one that has been prosecuted in its simplest forms, it is, therefore, otherwise than an art which, of all others, perhaps, affords the most varied scope and the largest sphere of development to the powers of the human mind. And it is most essential, if indeed it be true, as true it is, that a large part of the national welfare hangs upon its prosperity,—it is most essential that you should have the best and the most efficacious means of comparing its state in one year with its state in another, of recording for future encouragement the progress that has been achieved in the past ; and if perchance a time should come when in any one of its branches some partial failure should be perceived, that that failure should be noted at the first moment when it becomes visible, in order that the sense of the defeat may lead to its being at once repaired. . . . . Well, again, my lord, I will venture to give another reason why I myself, an uninstructed person, may venture to feel a sentiment of gratitude to those who, in this matter, give us the benefit of their instruction. If we look to the trade of the farmer, it seems to me to stand distinguished from all other trades,—not in the less, but in the greater amount of the demand that it makes on his mental powers. In point of fact, if we are to regard the farmer as an isolated man, he has got to struggle with everything. He ought to understand something of the universe in which he lives, and more or less of many sciences that fall within the range of human intellect. He ought to be profound in meteorology ; he ought to be a consummate chemist ; he ought to have a knowledge of birds and animals, as scarcely a lifetime could acquire. He ought to be a mechanist of the first order ; and, in point of fact, there is no end to the accomplishments which the individual farmer, to be a good farmer, if he stand alone, ought to possess. And if I take the case of two men setting out in life with a moderate capital at command—say, two men who have £5,000 each to dispose of, and the question being whether they are to enter into an ordinary trade, or whether they are to enter into the business of farming, I say that the man who takes £5,000 to stock a farm, which is let to him as a tenant farmer, will require far more intelligence in order to enable him to properly transact his business than if he opeu a shop in some street in a great city."

*   *   *   *   *   *

www.ingramcontent.com/pod-product-compliance
Lightning Source LLC
Chambersburg PA
CBHW021451090426
42739CB00009B/1718